TABLE OF CONTENTS

Prologue

This book covers the things that mining companies, mining executives, mining professionals and all expatriate personnel operating or planning a project in Africa need to know. I should stress that it does not reflect the realities of all African countries.

The lessons learned are based on my personal experience working for over four years in Africa as a Supervisor, Superintendent, Leader and Director in the mining industry, and setting up a company of my own. This is practical advice on matters that the experts do not tell you about, and you will not find it anywhere else; I have done the research and reported on it for your information. This is valuable information that never makes it to the board room, but if you follow it, it could make your operation and project a success story.

I was an expatriate working in the place where I was born (West Africa). I had the advantage of speaking the official languages of the countries I worked in. What I lacked was the knowledge that I needed to prepare me for working in Africa. All I had, coming out of the Pilbara where I had worked for one of the biggest mining companies in the world, was my expertise as electrical engineer and my leadership skills, which I have developed throughout my life. However, nothing prepared me for what I found when I arrived on site in West Africa. For the first time my first thought was, what I have I let myself in for? But there was no turning back; I had to make a go of it. This account is a compressed version of the things to do and those to avoid to make your project in Africa viable.

There are always two sides to a story; this side is mainly aimed at companies and expatriates working in Africa. It is what you really need to know when you are operating in Africa. As an expatriate and an African, I was often caught in the middle between the views of the expatriates and what they thought of the local staff and colleagues, and what local staff thought of the expatriate community working on their land. Being a very good listener and not taking sides (there was no side to take, as it is one team in my view) people just opened their hearts to me. I also observed operational decisions and actions that could have been done better to save money, and to avoid local staff striking and local communities blocking operations.

I have not mentioned any specific mining companies, operations and projects I have worked for in this book. However, the recommendations are based on my personal experience working in Africa, observing small details missed by most people. You now have something to help you to avoid common mistakes when operating in Africa which could potentially derail your project or operation at any time, damage your reputation and negatively affect people's lives. As a mining company or a specialist in your field, you know what to do get your specific commodity out of the ground in order to deliver to your customer. Therefore I am not discussing factors such as the economic viability of a project, and other factors associated with a mining project or operation which I assume you are good at doing. I'm giving you very important/useful details in order for you to succeed in Africa. Maintaining your Social License to Operate, and managing local communities' expectations are the challenges faced by mining companies in Africa in order to run a profitable operation where everyone is a winner.

Acknowledgments

This book was made possible through the support of outstanding people; without them this electronic book will not have been possible.

Dr Martin Soh my mentor and friend who has been there in good and bad times, he was my toughest critic; I was delighted by his positive feedback on: Mining in Africa The Important Things You Need To Know. Dr C H Houen, for her excellent editing skills and professional advice. Finally thank you to Adama Barry for her outstanding support thorough this project.

Chapter 1: The Myth That Mining Companies Have Unlimited Wealth at Their Disposal

One of the first things that I noticed when I got to my first site in Africa was that there was a myth that mining companies have unlimited wealth. This is probably unknown to most mining companies. But from my personal experience that belief was well established with the local communities and some of the local staff members I interacted with.

This myth was unwittingly reinforced by the expatriate communities themselves. For example, in one night an expatriate could spend more on alcohol than the monthly wage earned by local employees. For some of the unskilled workers, that one night of alcohol consumption could feed their entire family for more than a month, because his or her wage was less than that. In case you are wondering what is wrong with drinking as much alcohol as you want when you have a Rostered Day Off (RDO) the next day, here is the answer. You are probably in a poor country to do a specific job, not at a holiday resort where you can do as you please without anyone caring and where your actions or behavior will not affect anyone. At work and outside work hours all your actions are observed and scrutinized. You are seen as the professional/expert, and all your behavior is recorded by your local staff without you knowing it. You are doing more harm to your company than good by buying plenty of alcohol out of your own pocket for your team members or colleagues than by not doing so. No-one likes to see his/her colleague or boss intoxicated by alcohol, and it is worse in a poor country. I am not criticizing the action and behavior of drinking large amounts of alcohol; I am just telling you your local staff's point of view is: "why can't I get a small pay raise for the benefit of my family when expatriates are paid so much that they can spend more than my monthly wage in one night?" My advices to you is get your job done, keep a low profile outside work hours, and do as you please when you are on vacation outside the African country in which you are operating. A better way of doing things if you really want to be nice, instead of buying large amounts of alcohol for everyone, is to liaise with your community department and find a local orphanage or school and contribute to their development in your humble way.

Ensure that as an expatriate you do not brag about how much you are earning or everywhere you have been on your time off or the latest toy you purchase to your local staff (I have seen it happen). And If you are doing your job, keeping a low profile, and one of your local colleagues somehow finds out how much you are earning and questions why he or she is not earning the same wage, and sees you as rich and earning too much money — explain to him or her that in most developed countries, for a certain pay bracket more than 40% of your income gets taken as tax before you receive it; explain that the cost of living in the modern world is far greater than in any under-developed country. And that the perception of wealth for all people living in the modern world is only an illusion as most people are in debt. An example I can give is that an expatriate came up with a brilliant idea (so it was thought at the time) to make a "hall of fame" on the kitchen wall which consisted of each employee putting their cars on the wall. Only expatriates participated in the project; it went from the funniest/most beat up cars to some of the coolest and most expensive automobiles on the planet. One of the local staff did not find it amusing to see expatriates displaying their apparent "wealth" and made a complaint; so the "hall of fame" wall had to come down, creating more separation and distance between expatriate and local staff. Not to mention fuelling the myth that mining companies have unlimited wealth but only for the benefit of citizens of specific countries.

If everyone travels by road to go to site, ensure that this also applies to the CEO of your company when he or she gets to site. If it is safe for all your employees to travel by road, it is it is also safe for the CEO to do so. This reinforces the message that you have limited resources, and hiring a special plane or helicopter when the CEO arrives is counterproductive and generates waste. If you are a CEO and are reading this, ensure that you used the same mode of transportation as your team to get to your site.

For residential houses, set specific criteria for employees by limiting what accommodation they can have as a family or single person. Giving free choices is good but cannot always be the best image for a company with limited resources; for instance, if you have a family of four living in a house with 17 air conditioner units inside the house powered by a generator 24 hours a day.

Reduce your overheads: don't have more layers of management or staff than can be justified to achieve your goals. If you do not have a clear scope of work (different from a role description) as to why you hire a person and what you want her/him to achieve, do not hire her/him. Just because you have the funding, approval and a role description does not justify for you in hiring someone, as it generates waste for your company and confusion for the person as to why he or she has been hired; he or she will rightfully spend most of their time on social networks and add to the myth that mining companies have unlimited funds.

If you are not operational (not making any money) do not have two separate organizations to run your project, such as a construction project team with a Construction Managing Director and an operation team with an Operation Managing Director. If you do this you will be creating duplication for the same services in both organizations and confusion as to who is supposed to do what. Communication will not be done at the ground level, as it would have to go up the ladder of one organization before being communicated to the next organization if the information is to reach the employees it's aimed at. Direct communication at the lower level where the work is done will be impeded. This kind of structure at an early stage of your project will generate excuses for why work is not getting done; each organization will be blaming the other for why things are not happening, with no-one being accountable for the failure. You will find scapegoats instead of sharing responsibility for failures. In case of success there will always be shared responsibilities, even from the organization which did not pull its weight. Having one Managing Director responsible for the execution and all aspects of your project, with everyone reporting to him or her, will remove any confusion and save money in the process. This will remove silo communication, for communication will happen directly at the lower level where the job is actually done; for example, engineers will talk directly to community specialists and vice versa. Everyone will be working toward the same goal with no competing agenda, and you will be showing you are a company with limited resources which you are spending wisely.

When I worked in Australia, I noticed that some mining stores were equipped with good security systems such as CCTV and electronic tagging of items. However I did not notice the same level of security in any of the mining stores I have seen in Africa. Sometimes the quantity of the items on the electronic data base did not match the reality, giving the sense that mining companies have unlimited wealth, and that they cannot account for where all their equipment is as they can just order new stock.

Do not create inflation by paying above the average for the same good and services that everyone is paying for in the country. Do your research. If you don't you will be the fat cow that everyone is trying to milk.

Treat your employees and your local communities as your partners and lead by example.

If you do not follow this practical advice, do not be surprised to find your employees or local communities striking and demanding more from you as they have been observing and watching you. The fact that your local staff are paid above the national average in the country in which you operate does not make you immune from industrial actions. Managing your resources efficiently and paying attention to your personal conduct will make the difference.

Below are some of the guidelines for paying attention to details when working in Africa.

Planning

Most mining companies I have observed have a high level plan for how they will operate and execute their work. They sometimes get into trouble in Africa when it comes to execution. Most of the time it is because their Work Breakdown Structure (WBS) or details plan is based on what they did in the past; or they simply have not taken into consideration certain factors based on the specific realities of the country in which they are operating. Each Item on your WBS structure must have its own WBS; if you don't plan to this level of detail, you may get stuck and have unforeseen delays which may cost you so much money that your project will no longer be viable, given your resources. As a general rule, if your plans are mainly based on assumption and not on facts, you will run into trouble. At the early stage of planning your project, Involve qualified personnel with hands-on experience in the country you are operating. You will get into trouble if your plan is done only from the comfort of an office without thorough input from the eyes on the ground, and factoring all the details.

Procurement

You should know and verify how long it takes to bring items to your site, including the time it will take to clear customs. This includes sourcing your equipment as per your design specifications from the country where you are operating or the nearest country, instead of the country where your engineering team is based, which may be easy for the design engineers but could be an expensive exercise to implement during construction and operation. I recall that a General Manager got frustrated when I could not give him an answer as to when I would complete a capital project after three months of being on site. The reason was simple: I had no information as to when critical items that were ordered would get to site (the procurement team did not know). I was correct not to give him an estimation out of thin air, as my order arrived a year later. Things got better after that, but my point is: *do not "guesstimate" under any circumstances or pressure*. If you do not know the answer and you "guesstimate" you may be mobilizing personnel to the site prematurely and spending all your resources (because you have to pay staff when you hire them regardless) without the project being delivered. This applies to anything linked to your project; if you do not know how long a critical part of your project will take to achieve, do not make any assumptions until you have a clear answer. Let me make it loud and clear: NEVER GUESSTIMATE.

Customs Clearance

If you are not confident that you will manage customs clearance effectively, or are not confident of the capability of the Customs office to cope with the volume of the work, you will be putting them under pressure; have a contingency plan in place. This could include hiring an experienced and reputable third party with experience in the country in which you are operating to do your customs clearance on your behalf, with a clear scope of work and Key Performance Indicators included in your contract agreement.

Logistics

Find out if the country roads are accessible in order for you to transport all your equipment and personnel to site. If that is not the case have a detailed plan and involve the relevant country authority to map the route, especially if they are National roads.

Legal License and Permit to Operate/Construct

It can be a nightmare sometimes for mining companies to find out what sort of license, permit and other authorization they need to operate or build. It may take too long to get the necessary authorization. This is sometimes but not always due to lack of information from the host country, language barriers, and discussions which are usually conducted at the high level of management only (Senior Government officials and the Mining Executives/Senior Management). In my personal experience, if you take a trip to the relevant department with the assistance of your community/permit team, you will find all the relevant information you need readily available and people willing to assist you to get your authorization. I am usually courteous to people and never had any problem getting anything done with the local administration. I should warn you: do not pay any type of bribe to get any kind of authorization. It is counter-productive, it will come back and bite you when you least expect it, and it is illegal. Do not mobilize your workforce until you have all the necessary authorization. Based on experience, do not rely on the assumption that you will obtain the necessary authorization on time, regardless of whom the assurance came from.

Design

Design something simple that will be easy to maintain by your local workforce after your construction and commissioning team have left. Again, all your equipment as per your design criteria should be sourced from the country in which you are operating or the nearest one.

Local Human Resources

If you are operating in a country that is run down, chances are that the quality of your personnel will be reflecting the image of the country, with some exceptions. You will have to invest in training your local staff to raise the level to the quality that you will need to operate at. Ensure that you and your contractors give priority to unskilled workers from communities directly impacted by your project.

Time Does Not Necessary Equal Money

The modern world philosophy that time equals money does not usually apply to some African countries; if it did probably all countries in Africa would have become developed countries. When setting your schedule you must have a detailed plan for how you will get from point A to point B with all the resources and qualified supervision needed to achieve your desired outcome. If you do not detail your plan enough you will fail.

Health, Safety and Environment

The conditions of the country you operate in, such as road traffic, waste disposal, safety on construction sites, and the health care system, will give you an indication of some of your local personnel needs for Health, Safety and Environment.

You have two choices. You can have a glossy "copy and paste" policy towards Health, Safety and Environment (HSE), turn a blind eye to what is actually happening, and act surprised and react when things go wrong, blaming personnel for not following policies and procedures. The better choice is to actually write your own HSE policies and standards that are practical and the reflection of what is occurring on your operation, in order to create an HSE workplace free of hazard, and lead by example by setting a strong culture of Health, Safety and Environment. Be proactive rather than reactive. Don't only talk about HSE, create an environment where you as well as everyone in your company will want to work.

I am not an expert in the medical field, so everything I mention in this paragraph should not be taken as expert advice, but as an observation. If you have a health condition that needs special attention, speak to your General Practitioner about it before leaving for Africa. Keep a healthy lifestyle, as the level of stress you may be put under could be worse than what you have ever experienced. Mining companies usually have good health facilities and medical staff on site. However, your special health needs may not be catered for on-site or in the country in which you are operating, should you run into problems. Public health care varies from one African country to the next; in some you will get the same level of care you will get in any developed country, while in others you will hope that you do not get sick. I once assisted a presentation regarding malaria, where the medical doctor mentioned that the malaria prevention tablet does not necessarily prevent you from getting malaria, but prevents you from contracting the worst form of malaria. (I did not verify this information but offer it so you know what to ask). Do not take any chances, stay indoors if you can after dark, and sleep under a mosquito net without any holes in it, or at least wear long pants, long sleeves and enclosed shoes if you are out after dark, to reduce the chance of being bitten by mosquitoes and contracting malaria. Depending on who you work for, some companies have a great management system to prevent all their staff from contracting malaria, while with others you will just do as you please.

Construction and Commissioning

Verify that you will be able to work safely during the rainy season, as this could impact your schedule (including road access). Don't just rely on data, as certain areas can remain flooded for prolonged periods of time after the rain has ceased. Ensure that your operation readiness team is part of the commissioning team, including your local staff. Verify that you have "as built" drawings after commissioning the construction, and not drawings issued before construction, which do not reflect what has been built. Check that the project has been delivered as per the design specification. Depending on the country in which you are operating, you may have to raise the standard of your local contractors, as they may not deliver to your specification and as per your schedule. In case you are wondering why I am stating the obvious, it is because I have witnessed everything I mentioned even in operating plants in Africa.

Social License to Operate

Prior to mobilization and when your team is on the ground, ensure that you have and will maintain your social license to operate throughout your project life cycle (see next chapter).

Chapter 3: Social License to Operate

Getting your social license to operate (SLO) will make your project viable for its life cycle. You can do everything right, but if you do not get your community's approval and acceptance right in order to set performance standards for your SLO, you will not have a project, or your operation will come to a halt when you least expect it. This chapter will provide you with information available in the public domain that some of you may not even know exists. I did not know this information (SLO) I am about to share with you existed, until I left the engineering department and went to work in the community department and that was after being in Africa for more than two years. I took my time to select some of the best performance standards for communities and social performance in order for you to appreciate the level of work you need to do to get and maintain your SLO.

To really appreciate local communities' issues, you have to put yourself in the shoes of local communities in which you are operating. Let's assume that there is a mining project initiated by a foreign company in your suburb. This project could mean that you will have to be resettled. Supposed that in your suburb there is cultural heritage; for example a cemetery with war heroes' final resting place. With that in mind you will be better prepared when dealing with local communities. Any idea that local communities living on their lands are just poor people who can just be paid off or pushed over by the might of a mining company with the support of a government to access their land is wrong and misleading.

The International Finance Corporation (IFC, http://www.ifc.org/) and other organizations have gone to great lengths to provide community performance standards and handbooks to ensure that you have the right tools to deal with local communities for your project's life cycle. The list of performance standards I have selected for you is not exhaustive, but would give you the minimum community and social performance standards to follow. Putting these standards into practice will assist into getting and maintaining your SLO in Africa and anywhere else in the world where there are local communities.

Social and Environment Impact Assessment: IFC Performance Standard 1: Assessment and Management of Environmental and Social Risks and Impacts

The objectives of IFC Performance Standard 1 are to:

a)   To identify and evaluate the environmental and social risks and impacts of the project.
b)   To adopt a mitigation hierarchy to anticipate and avoid or minimize residual impact on workers, affected communities and the environment.
c)   To promote improved environmental and social performance of clients through the effective use of management systems.
d)   To assure that grievances from affected communities and external communication from other stakeholders are responded to and managed appropriately.
e)   To promote and provide means for adequate engagement with affected communities throughout the project life-cycle issues that could potentially affect them, and ensure that relevant environmental and social information is disclosed and disseminated.

The requirements for IFC performance standard 1 are to have an Environmental and Social Assessment and Management System which will incorporate the following elements:

a) Policy
b) Identification of risks and impacts
c) Management programs
d) Organizational capacity and competency
e) Emergency preparedness and response
f) Stakeholder engagement
g) Monitoring and review.

Community Health, Safety and Security: IFC Performance Standard 4: Community Health, Safety and Security

The requirement for Performance Standard 4 is to:

a)   Evaluate the risks and the impacts to the health and safety of the affected communities during the project life-cycle.
b)   Establish preventive and control measures consistent with good international industrial practice.
c)   Identify risk and impacts and propose mitigation measures that are equal in nature and magnitude. This measure would be used to avoid risk and minimize impacts.

Resettlement Action Plan: IFC Performance Standard 5: Land Acquisition and Involuntary Resettlement

The objectives of IFC Performance Standard 5 are to:

a) Avoid and minimize displacement by exploring alternative project designs.
b) Avoid forced eviction.
c) Anticipate, avoid and minimize adverse social and economic impacts from land acquisition or restrictions on land use by:
   1) Providing compensation for loss of assets at a replacement cost.
   2) Ensuring that resettlement activities are implemented with appropriate disclosure of information, consultation, and the informed participation of the affected local communities.
d) Improve or restore the livelihoods and standards of living of displaced persons.
e) Improve living conditions among physically displaced persons through the provision of adequate housing with security of tenure at resettlement sites.

The requirement for Performance Standard 5 is to have:

a) Feasible alternative project designs to avoid or minimize physical and/or economic displacement, while balancing environmental, social, and financial costs and benefits, paying particular attention to impacts on the poor and vulnerable.
b) Compensation and benefits for displaced persons.
c) Community engagement.
d) Grievance mechanisms.
e) Resettlement and livelihood restoration planning and implementation.

Indigenous Population: IFC Performance Standard 7: Indigenous People

The objectives of IFC Performance Standard 7 are to:

a) Ensure that the development process fosters full respect for the human rights, dignity, aspirations, culture, and natural resource-based livelihoods of Indigenous people.
b) Anticipate and avoid adverse impacts of projects on communities of Indigenous people and minimize and/or compensate for such impacts.

c) Promote sustainable development benefits and opportunities for Indigenous people in a culturally appropriate manner.

d) Establish and maintain an ongoing relationship based on Informed Consultation and Participation (ICP) with the Indigenous people affected by a project throughout the project's life-cycle.

e) Ensure the free and informed consent of the affected communities of Indigenous people.

f) Respect and preserve the culture, knowledge, and practices of Indigenous people.

The requirement of IFC Performance Standard 7 is to avoid adverse impacts, and to have participation and consent from Indigenous people.

Cultural Heritage: IFC Performance Standard 8, Cultural Heritage

The objectives of IFC Performance Standard 8 are to:

a) Protect cultural heritage from the adverse impacts of project activities and support its preservation.

b) Promote the equitable sharing of benefits from the use of cultural heritage.

The requirement for IFC Performance Standard 8 is for the Protection of Cultural Heritage in Project Design and Execution.

Here are some very important handbooks that will assist you in setting up your own policies and procedures in order for you to obtain and maintain your Social License to Operate:

Community Consultation and Engagement: IFC Handbook: Stakeholder Engagement: A Good Practice Handbook For Companies Doing Business in Emerging Markets

PART ONE of the IFC Handbook discusses:

Key Concepts and Principles of Stakeholder Engagement

a) Stakeholder identification and analysis
b) Information disclosure
c) Stakeholder consultation
d) Five steps for iterative consultation
e) Informed participation

f) Consultation with Indigenous people

g) Gender considerations in consultation

h) Negotiation and partnership

i) Grievance management

j) Stakeholder involvement in project monitoring

k) Reporting to stakeholders

l) Management functions

PART TWO of IFC Handbook discusses:

Integrating stakeholder engagement with the project life cycle

a) Project concept

b) Feasibility studies and project planning

c) Construction

d) Operations

  e) Downsizing, decommissioning, and divestment

Projects Induced in-migration (IFC Handbook: Projects and People: A Handbook for Addressing Project-Induced In-Migration)[1]

The business case for addressing project-induced in-migration

Describes the main impacts of project-induced in-migration, and the trade-offs between proactive and reactive management of its associated impacts.

The project-induced in-migration phenomenon

Considers the dynamics of project-induced in-migration, and its potential environmental and social impacts. It considers the unique aspects of in-migration in relation to artisanal and small-scale mining, resettlement, indigenous peoples, areas of high biodiversity value and cultural heritage.

The assessment of the probability of project-induced in-migration and the risks that such in-migration poses to a project

The basic tools to assess the probability of and the risks posed by in-migration, and the requirements for a situation analysis as a basis for developing a project-specific influx-management strategy and plan.

Management approaches

A comprehensive description of potential management approaches, including approaches to reducing in-migration, managing its footprint, enhancing its positive impacts, preventing and mitigating its negative impacts. Interventions supporting each of these approaches are described.

Developing a management strategy and integrating it into the project

This part will assist in the development of a strategy and plan.

The handbook has document templates and a case study on the Cameron Chad pipeline.

----

[1] Project-induced in-migration (or influx) involves the movement of people into an area in anticipation of, or in response to, economic opportunities associated with the development and/or operation of a new project. http://commdev.org/migration-0

Community Complaints, Disputes and Grievances: IFC Guidance Note, Good Practice Note, Addressing Grievances From Project Affected Communities, Guidance For Projects And Companies On Designing Grievances Mechanism.

The handbook discusses the basic elements of grievance mechanism design:

- It discusses the project-level grievance mechanism, and why Is It needed;
- It defines what is a grievance;
- It defines what is a project-level grievance mechanism;
- It defines who will use a project-level grievance mechanism;
- It defines how a grievance mechanism benefits companies and communities.

Part I: Describes the principles of a good grievance mechanism:

a) Principle 1: Proportionality: a mechanism scaled to risk and adverse impact on affected communities.
b) Principle 2: Cultural appropriateness: designed to take into account culturally appropriate ways of handling community concerns.
c) Principle 3: Accessibility: a clear and understandable mechanism that is accessible to all segments of the affected communities at no cost.
d) Principle 4: Transparency and accountability to all stakeholders.
e) Principle 5: Appropriate protection: a mechanism that prevents retribution and does not impede access to other remedies.

Part II: Describes the process steps for grievance management:

a) Step 1: Publicizing grievance management procedures.
b) Step 2: Receiving and keeping track of grievances.
c) Step 3: Reviewing and investigating grievances.
d) Step 4: Developing resolution options and preparing a response.
e) Step 5: Monitoring, reporting, and evaluating a grievance mechanism.

Part III: Discusses management of a grievance mechanism and what resources are needed:

a) Resources for grievance mechanisms.
b) Who should be responsible for implementation?
c) Discuss if your internal capacity is sufficient?
d) When should third parties be involved?
   e)   Are grievance mechanisms needed for projects implemented by contractors?

Security and Human Rights: The International Council on Mining and Metal: Voluntary Principles on Security and Human Rights: Implementation Guidance Tools (https://www.icmm.com/)

This publication discussed the following:

a) Understanding the voluntary principles (VPs)
b) Stakeholder engagement
c) Risk assessment
d) Public security providers
  e) Private security providers.

The Appendices discuss the following:

a) Voluntary principles on security and human rights
b) Human rights articles and their relevance to the voluntary principles case studies
c) Stakeholder mapping tool
d) Worked example of risk assessment
e) Risk assessment information sources
f) Interacting with public security
g) Use of force and firearms
h) Rules of engagement
i) Use of force
j) Firearms procedures

k) Private security and community engagement
l)   Sample contract clauses on VPs for private security contracts
m) Incident reporting
   n) Sample service level agreement.

Although this may look like a lot of work and effort, it's important to get this right to obtain and maintain your Social License to Operate. I will give you a tip based on my experience on writing standards. Instead of writing different sets of policies addressing each term of reference, write one standard addressing each term of reference in your own words. For example, two terms of reference that are used throughout are stakeholder engagement and grievance mechanisms. So when you are writing your standards you will be writing in your own words, based on your specific reality and on the information, so that you now have one standard for addressing stakeholder engagement and grievance mechanisms. Regardless of the activities you are doing, it should be the same process and consistent. Do not have one policy for stakeholder engagement when you are doing resettlement activities and another policy for stakeholder engagement when you are doing impact assessment; that will confuse everyone, including your own staff. If for your specific project, Indigenous population does not apply, there will be no need for you to include it in your standards and you can specify that this does not apply to your project/population. One more tip: do not include your data or procedures in your standards, but you can reference them as separate documents.

In the mining industry, there is a saying that safety is a priority and it is everyone's business. To operate successfully in Africa or anywhere in the world you should have the same passion for local community issues as you do for safety. There was a time in the mining industry when safety was not that important, but now almost every mining company realizes that not only is safety important, it is good for business. The same goes for community investment: it is good for your business to get it right. This is not charity, it is a sound business investment and being a good corporate citizen. Communities and social performance should not be left as the thing the community department does, and should be everyone's business, involving all departments and contractors, with everyone knowing and following your Community and Social Performance policies. By that I do not refer to your corporate or the IFC communities' standards, but I refer to your site or project specific Community and Social Performance standards written in your own words. If your Senior Management team, especially your Managing Director and your chief Financial Officer are not leading your communities' performance standards, you will most likely have trouble obtaining your Social License to Operate.

The performance standards and handbooks referred to in this chapter are freely available on the IFC website and the International Council on Mining and Metal. It is good for your business to invest in communities and social performance. Getting it wrong will cost you your Social License to Operate, money in lost revenue, damage to your reputation, and can potentially damage the lives of many people.

Chapter 4: Develop or Hire Leaders

Developing or hiring leaders for your organization is important if any organization is to do well. In Africa it is a must to have leaders running your company in order to be successful. If you have leaders running your organization, I will be surprised to see industrial trouble. Good leaders will obtain and maintain your Social License to Operate throughout the life cycle of your project.

I have seen top managers struggling in Africa, where things do not happen the way you want, or as they may do in developed countries where the rules and regulations are followed by most employees, and where your contractors know what to do and deliver as per your specifications and schedule, and if managed effectively, on budget. In parts of Africa where I have worked, the game is different. You have to think outside the box; the unexpected can happen at any time, and you may find problems that you were not prepared for despite your years of experience, and to which you have to respond promptly without damaging the reputation of your company or escalating the problems. At time things could appear chaotic and it may appear that you are not making any progress. You will feel frustrated, and wonder — what am I doing here? In this kind of environment, managers struggle and leaders thrive. Therefore, to succeed it is best to hire or develop leaders who will inspire and drive the workforce to achieve the desired outcome. They will adopt and implement the community policies in order for you to have and maintain your Social License to Operate. Where a Manager will find reasons for not delivering results, a Leader will thrive on a challenge and will have a clear picture and the resources needed to achieve success and bring his or her team to the desired outcome. A leader will not be afraid to make decisions and will be the first to put his or her hand up when things do not go according to plan, instead of finding someone else to blame.

To be a leader, one needs to have the capability to produce outstanding positive results and changes to an organization. To produce these changes, the leader must have the necessary skills and enthusiasm to motivate the people within the organization to implement all the changes and achieve the goals set by the leader.

For a leader to successfully create a level of confidence within his/her group members, it is essential that they trust the leader first. An important trait that an efficient leader displays is trustworthiness and honesty.

Leaders are often outgoing, but it is not essential to display this trait to be an effective leader. Despite being outgoing or reserved, a leader must be assertive. Assertiveness is to express his or her opinion freely and convincingly without offending his or her subordinate. Positive emotional energy is the key to a successful leadership role, and brings with it happiness and well-being. When a leader is positive in their approach then their team will achieve better results in all areas. Another approach a positive leader needs is to be able to keep calm in any situation. Leaders are more relaxed, more gentle, and more in control of their emotions than the average person. They are very aware that expressions of negative emotions can deprive them of the energy they need to be effective in the work place. They do not allow themselves to become upset or angry over little unimportant issues or even over larger issues.

A trait that Leaders have is high emotional intelligence, which can be defined as the ability to understand one's feelings and those of others. It is separate from academic intelligence, which is purely cognitive. Many people with high IQs and lower emotional intelligence often find themselves working for people with a lower IQ but with a higher emotional intelligence.

Most people don't know the difference between management and leadership. "Broadly speaking, leadership deals with the interpersonal aspects of a manager's job, whereas planning, organizing, and controlling deal with the administrative aspects. Leadership deals with the change, inspiration, motivation and influence" (Dubrin & Dalghlish, 2003, p. 4).

<u>**Leader Vs Manager**</u>

Leader	Manager
• Visionary	• Rational
• Passionate	• Consulting
• Creative	• Persistent
• Flexible	• Problem solving
• Inspiring	• Tough-minded
• Innovative	• Analytical
• Courageous	• Structured
• Imaginative	• Deliberative
• Experimental	• Authoritative
• Independent	• Stabilizing
• Shares knowledge	• Centralizes knowledge

Reference:

Dubrin, A. & Dalglish, C. (2003). Leadership: an Australasian focus. Australia, Queensland: John Wiley & Sons Australia.

Chapter 5: Power Requirement

If the country's utility provider cannot provide reliable power to the country, chances are that they will not provide you with power even after you have invested in power utilities and have a contract in place stating they will provide you with power to your specifications and sustain penalties if power is not delivered. If the power provider has not been able to provide reliable power to the country for years, despite their best intentions, nine times out of ten you can be sure that they will not provide you with reliable power. If you decide to invest in local power utilities in order to hand over to the local power authority that has not been able to supply power to its own country for years, I suggest that you take your investment and throw it in a local river, as someone may find it downstream and put it to a better use. (Don't do that either, it is a metaphor!). However, what I will recommend is that, with your investment in the local power utility, you should also have a plan to take on the management of the power plant you have invested in and the training requirements of the staff, until such time as you are satisfied that the performance level has risen to your satisfaction and you can gradually hand over the management to the local power authority.

Mining companies are very conservative and like to do what they have been doing for decades. However, start being innovative by reducing your operating costs. Look for reliable and constant sources of energy as an alternative to diesel power. You may need to think about creating a new power supply chain for your project, or think of hybrid systems consisting of renewable and conventional energy, which could be proven to be cheaper for the life cycle of your project and good for the environment.

Verify that your memos are not interpreted as socially insensitive. Values differ from one country to the next, and what is a normal statement in a country could be seen as an offence in another country. In order to be sure, have all your memos overlooked by one of your senior local staff who is versed in local tradition.

Managing confidential information will need to be treated carefully, as all your confidential information may find its way to the local news and could be distorted. It is important to discuss what information you are sending and how it will be perceived if it leaks out.

Some of your local staff will be inclined to communicate verbally rather than by email, and others do not have access to corporate email. It is important to ensure that important messages have been passed on verbally so that your messages reach all personnel.

Remember you are in a foreign country and if the official language is not the same as yours, I advise that if you are a Supervisor, Superintendent, Manager or Senior Manager on a long term assignment, you should learn the official language of the country or state you are in, and ask your company to provide you with access to tuition. This may seem an additional burden at first, but you will find long term benefit in doing this. Just as you would like a foreigner working in or migrating to your own country to learn the official language, so too, you will demonstrate your commitment and respect for the country in which you are operating by learning their language. This could also save you lots of trouble, because sometimes the message you are trying to convey (usually verbal) can get lost or distorted in translation.

Ensure that you have a literacy program so that all your local personnel will be able to read and write. Illiterate personnel are easily manipulated during industrial strikes, as your messages might never reach them. It would also be a challenge for you to effectively train personnel that are not literate in all your health, safety, environment and community procedures and standards.

Being confident in one country can be seen as being arrogant in another. If that is the case you will have a nightmare getting anything done. Ensure you learn the culture and greeting protocol in your host country. It is wise to refrain from making jokes, as a harmless joke in your culture can be an offense in another culture and vice versa.

Chapter 7: Corruption

If the corruption index in the country you are operating in is very high you will probably experience corruption in two different ways.

1) You will be approached to make or receive payments under the table for a service
2) You will sign for illegal transactions without you even being aware.

I have experienced number one on several occasions, and set up some measures so that number two cannot be done with my authorization. Everything I signed was very well documented, so that if it was an illegal transaction I could pinpoint who the culprit was. For example, I was signing for cement to be delivered on a specific project where I could not be on every single site to see if the item was being delivered (that is why you have supervisors), but I made sure that every team leader or supervisor making the request had to counter sign. Anyone making an illegal transaction through you will not want to leave any trace, so if the person has to countersign, they will think twice. So if an audit showed that less cement was used for specific projects than had been signed for, I could go back into my records to trace the potential person(s) who made illegal requests. That is just one example out of many. Before hiring or going to Africa, check that you have these values in place:

Honesty
Integrity
You meet the criteria for the job
You will treat your staff as you would like your boss to treat you.

In a country where the corruption index is high, corruption is not seen as such, but as how they do business. Note that not all your local employees will be tempted to engage in illegal/corrupt activities, just as not all expatriates are honest and working with integrity. How you tackle corruption within an organization is up to you; I am only making you aware of the reality, for no person or organization will be immune from it.

Do not pay any money under the table, or receive any money in return for a service; this includes giving money under the table to Government Officials to obtain a concession. If you do you will never succeed, and do not be surprised if one day you lose your License to Operate. Always work with honesty and integrity and set a very good example for your employees and the country in which you are operating.

Politics

Stay out of politics. Regardless of how badly you think a country is being run, do not give your honest opinion; a private conversation can find its way into the public domain, and you can get into lots of trouble. Stick to your mission of being a mining company manager or employee, and instead of criticizing anyone, ensure that your Social License to Operate is in good order. It is up to local people themselves to decide which direction they want their own country to go.

Do not make promises you cannot keep.

Emotion

Emotion will be part of life and doing business in Africa. You will have to strike a delicate balance between teaching your local staff to remain professional and your own need to learn and respect the culture of the country you are operating.

Site Accommodation

In an ideal world I would advise you to have the same accommodation for everyone. However, if you decide to have your site accommodation based on job titles, make sure that it applies to everyone. This means that expatriates do not get preference over local employees in violation of your on-site accommodation policy, as this may bring tension and division instead of one team.

Promotion of Your Local Staff

I strongly encourage you to promote your local staff into leadership and senior positions, however do not do so under pressure because you are trying to save money or the government is asking you to do so. What I advise you to do is to have clear key performance indicators for advancement of your local staff and provide them with the skill and training requirements to get there. If you promote them knowing that they are not ready, you will be setting them and your company up for failure. Not having goals or policies for the advancement of your local staff into leadership positions is also not advisable, and bad for your business.